MW00522871

SALINA LIBRARY
100 BELMONT STREET
MATTYDALE, NY 13211
315-454-4524

EARLY BIRD STORIES™

Let's Notice Forms of Water

Martha E. H. Rustad   Illustrated by Christine M. Schneider

LERNER PUBLICATIONS ◆ MINNEAPOLIS

NOTE TO EDUCATORS

Find text recall questions at the end of each chapter. Critical-thinking and text feature questions are available on page 23. These help young readers learn to think critically about the topic by using the text, text features, and illustrations.

Copyright © 2022 by Lerner Publishing Group, Inc.

All rights reserved. International copyright secured. No part of this book may be reproduced, stored in a retrieval system, or transmitted in any form or by any means—electronic, mechanical, photocopying, recording, or otherwise—without the prior written permission of Lerner Publishing Group, Inc., except for the inclusion of brief quotations in an acknowledged review.

Lerner Publications Company
An imprint of Lerner Publishing Group, Inc.
241 First Avenue North
Minneapolis, MN 55401 USA

For reading levels and more information, look up this title at www.lernerbooks.com.

Photos on p. 22 used with permission of: NASA/MSFC (Earth); Deacons docs/Shutterstock.com (ice melting); fizkes/Shutterstock.com (kids drinking water).

Main body text set in Billy Infant.
Typeface provided by SparkyType.

Library of Congress Cataloging-in-Publication Data

Names: Rustad, Martha E. H. (Martha Elizabeth Hillman), 1975- author. | Schneider, Christine, 1971- illustrator.
Title: Let's notice forms of water / Martha E. H. Rustad; illustrated by Christine M. Schneider.
Description: Minneapolis : Lerner Publications, [2022] | Includes bibliographical references and index. | Audience: Ages 5-8 | Audience: Grades K-1 | Summary: "Ms. Ling's students work as science detectives and make observations of water. Young readers will love exploring water in its solid, liquid, and gas forms in this entertaining, illustrated story"—Provided by publisher.
Identifiers: LCCN 2021012946 (print) | LCCN 2021012947 (ebook) | ISBN 9781728441382 (library binding) | ISBN 9781728444666 (ebook)
Subjects: LCSH: Water—Juvenile literature.
Classification: LCC GB662.3 .R86 2022 (print) | LCC GB662.3 (ebook) | DDC 551.48—dc23

LC record available at https://lccn.loc.gov/2021012946
LC ebook record available at https://lccn.loc.gov/2021012947

Manufactured in the United States of America
1-49909-49752-5/26/2021

TABLE OF CONTENTS

Chapter 1
Science Detectives.....4

Chapter 2
Missing Ice Mystery.....6

Chapter 3
Missing Puddle Mystery.....12

Chapter 4
Global Water.....16

Learn about Forms of Water....22

Think about Forms of Water:
Critical-Thinking and Text Feature Questions....23

Glossary....24

Learn More....24

Index....24

SCIENCE DETECTIVES

In Ms. Ling's class, we are going to be science detectives and observe **water**.

A detective is someone who solves mysteries. A mystery is something that is hard to explain or understand.

✓ Check! What will Ms. Ling's class observe?

CHAPTER 2

MISSING ICE MYSTERY

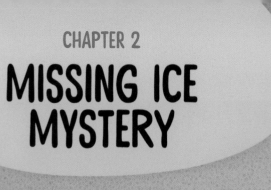

Ms. Ling put an ice cube in a cup this morning.

But where has the ice cube gone?

"The ice melted!" Lola says. "There's water in the cup."

Ms. Ling says, "Correct. Frozen water melts when it warms."

Frozen water is a solid. Melted water is a liquid.

When water gets cold, it changes from a **liquid** to a **solid**.

CHAPTER 3

MISSING PUDDLE MYSTERY

Ms. Ling says, "Yesterday I drew a circle around one puddle."

The **puddle** is missing!

"Did someone take it?"
Aleekah asks.

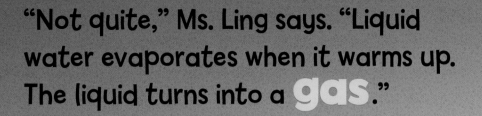

"Not quite," Ms. Ling says. "Liquid water evaporates when it warms up. The liquid turns into a **gas**."

The sun shined on the puddle. The puddle warmed up and evaporated.

✓ Check! Why did the puddle evaporate?

15

CHAPTER 4

GLOBAL WATER

We look at a globe to find water on Earth.

A globe is a round map of Earth.
Water is blue on it and land is green.

"The white part on the bottom and top of the globe are the South and North Poles," says Ms. Ling. "They are made of **ice** and **snow**."

18

"Ice and snow are made of water!" Becky adds.

Ms. Ling asks, "What did we learn about **water** today?"

"Water freezes when it gets cold," says Aleekah. "And ice melts when it gets warm."

gas   liquid   solid

"When water heats up, it turns into a gas," Malik adds.

Ms. Ling says, "Great work, science detectives!"

✓ Check! What are ice and snow made of?

LEARN ABOUT FORMS OF WATER

Frozen water begins to melt at 32°F (0°C) or warmer. It changes from a solid to a liquid.

Liquid water freezes at 32°F (0°C) or colder. It changes from a liquid to a solid.

Water in liquid and solid forms covers about three-fourths of Earth. That means water covers more of our planet than land does.

Almost all the water on Earth is salt water in oceans. Freshwater fills most lakes and rivers. Freshwater is not salty. Humans can only drink freshwater.

Evaporation happens when a liquid changes to a gas. The hotter the air temperature, the faster the liquid evaporates.

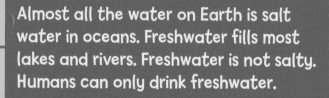

THINK ABOUT FORMS OF WATER:
CRITICAL-THINKING AND TEXT FEATURE QUESTIONS

Is rain a solid, a liquid, or a gas?

How much water do you drink in a day?

On what page will you find the table of contents?

Who illustrated this book?

LERNER **SOURCE™**

Expand learning beyond the printed book. Download free, complementary educational resources for this book from our website, www.lerneresource.com.

GLOSSARY

evaporate: to change from a liquid to a gas

gas: a form of matter that spreads out to fill any space, such as air

liquid: a form of matter that is wet and can be poured

observe: to look or watch something closely

solid: a form of matter that is firm or hard and holds its shape

LEARN MORE

Carlson-Berne, Emma. *Let's Explore the Water Cycle.* Minneapolis: Lerner Publications, 2022.

Spalding, Maddie. *Solids, Liquids, and Gases.* Mankato, MN: Child's World, 2020.

Water—NASA Climate Kids
https://climatekids.nasa.gov/menu/water/

INDEX

Earth, 16–19

evaporation, 14–15

gas, 14, 20–21

liquid, 9–10, 14, 20–21

solid, 9–10, 20–21